The Class Party

by Amy Ayers
Photographs by Kay McKinley

Copyright © Gareth Stevens, Inc. All rights reserved.

Developed for Harcourt, Inc., by Gareth Stevens, Inc.
This edition published by Harcourt, Inc., by agreement with Gareth Stevens, Inc. No part of this publication may be reproduced or transmitted in any form or by any means, electronic or mechanical, including photocopy, recording, or any information storage and retrieval system, without permission in writing from the copyright holder.

Requests for permission to make copies of any part of the work should be addressed to Permissions Department, Gareth Stevens, Inc., 330 West Olive Street, Suite 100, Milwaukee, Wisconsin 53212. Fax: 414-332-3567.

HARCOURT and the Harcourt Logo are trademarks of Harcourt, Inc., registered in the United States of America and/or other jurisdictions.

Printed in the United States of America

ISBN 13: 978-0-15-360170-5
ISBN 10: 0-15-360170-1

1 2 3 4 5 6 7 8 9 10 039 16 15 14 13 12 11 10 09 08 07

SCHOOL PUBLISHERS

Today is the last day of school. Miss Green's class is having a party. There are muffins and juice for the class to share.

The children will eat and play games together. They will hear a story. Then it will be time to leave. It will be a fun day!

7

First the children get their muffins. There are banana muffins and blueberry muffins. The muffins will be a good snack. There are 7 banana muffins to eat.

$7 - 5 = 2$

Miss Green gives the children 5 banana muffins from the table. How many banana muffins are left? There are 2 banana muffins left on the table.

8

Now the other children get their muffins. These muffins have tasty blueberries in them. There are 8 blueberry muffins to eat.

$8 - 4 = 4$

Miss Green gives the children 4 blueberry muffins from the table. How many blueberry muffins are left? There are 4 blueberry muffins left on the table.

3

The children choose their juice. There is grape juice. There is apple juice too. The juice will taste good with the muffins. There are 3 children who want grape juice to drink.

$3 + 7 = 10$

There are 7 more children who want grape juice to drink. There are 10 children in all who want grape juice. The grape juice tastes great.

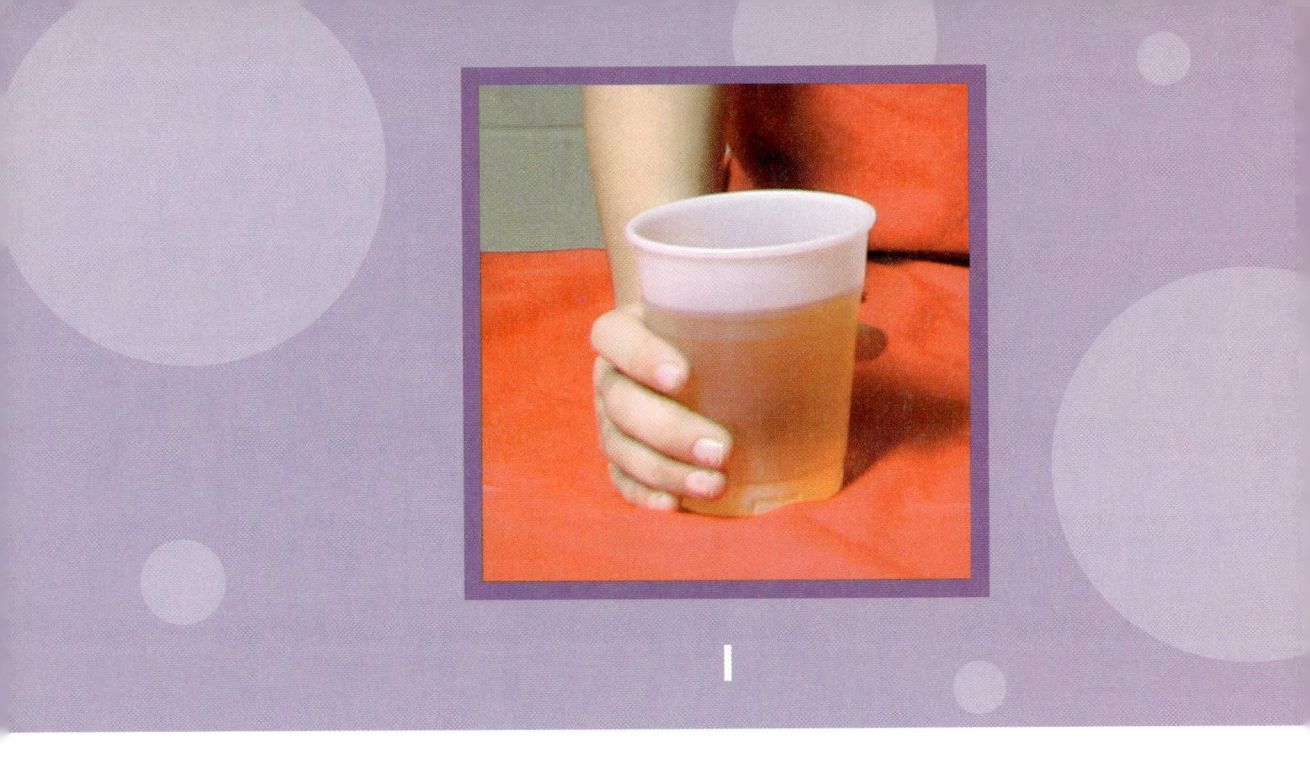

1

The children are almost done getting snacks. Some children still need juice. There is 1 child who wants apple juice to drink.

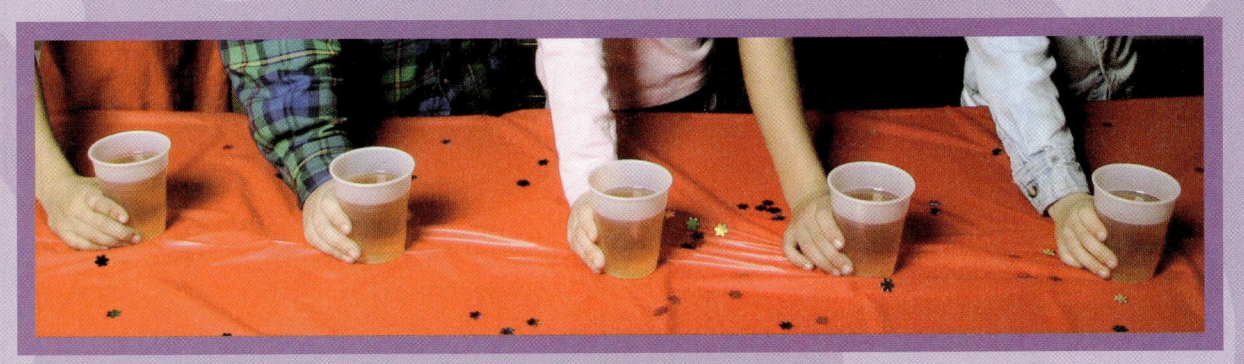

$1 + 4 = 5$

There are 4 more children who want apple juice to drink. There are 5 children in all who want to drink apple juice. The apple juice tastes sweet.

9

Snack time is over, and now the children can choose something to do. Some children want to paint. There are 9 paintbrushes for painting.

$9 - 5 = 4$

They take 5 paintbrushes from the table. How many paintbrushes are left? There are 4 paintbrushes left on the table. Now the children can paint pictures.

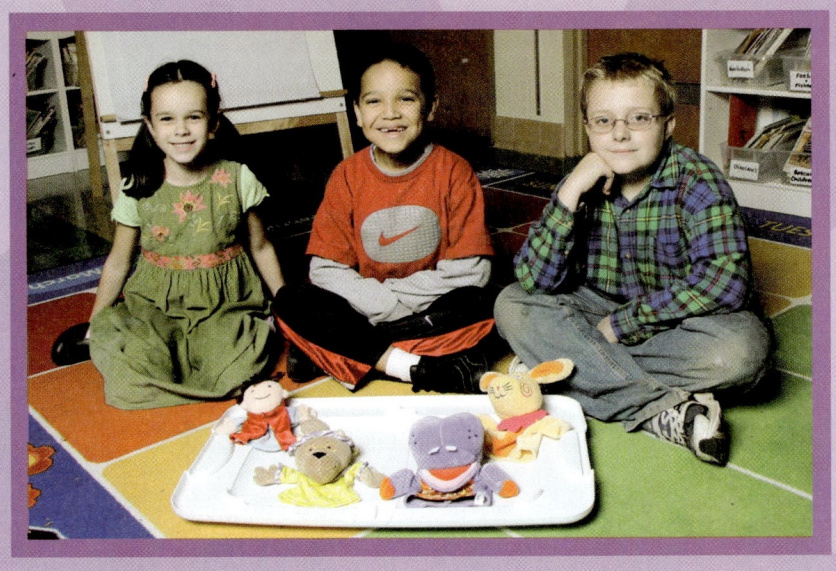

4

Some children want to play with puppets. They will give a puppet show for the class. The show will be about fun things to do during the summer. They have 4 puppets on a tray.

$4 - 3 = 1$

The children take 3 puppets from the tray.
How many puppets are left? There is
1 puppet left on the tray.

4

Some children want to play with wooden blocks. What will they build with the blocks? They will use the blocks to build a house. First they take out 4 big blocks.

4 + 2 = 6

Then the children take out 2 small blocks. Now they have 6 wooden blocks in all. They are ready to build the house.

10

Other children want to use blocks to make a pattern. They choose a pattern to make. They need 10 pattern blocks to make this pattern.

$10 - 7 = 3$

The children have only 7 pattern blocks. How many more pattern blocks do they need to make this pattern? They need 3 more pattern blocks.

9

Soon it is time to clean up and listen to Miss Green read a story. She will read a story about the last day of school. First 9 children sit down on the rug.

$9 + 6 = 15$

Then 6 more children sit down on the rug.
Now 15 children in all are sitting on the rug.
Miss Green reads the story.

The bell rings, and it is time to leave. The children say good-bye to Miss Green. She tells them to have a good summer.

It was a fun year for everyone. The children learned a lot and made many friends. They cannot wait to come back to school!

Glossary

equals = 1 + 7 = 8
1 plus 7 equals 8.

minus − 8 − 6 = 2
8 minus 6 equals 2.

muffin a small round cake made to serve one person

plus + 6 + 3 = 9
6 plus 3 equals 9.

puppet a doll that fits over the hand